Coenzyme Q10

The Essence of Energy

Barbara Wexler, MPH

Copyright © 2007 Barbara Wexler

All rights reserved. No part of this part of this publication may be reproduced, stored in a retrieval system, or transmitted in any form without the prior permission of the copyright owner.

For ordering information and bulk order discounts, contact:
Woodland Publishing, 448 East 800 North, Orem, UT 84097
Toll-free telephone: (800) 777-BOOK

Please visit our Web site: www.woodlandpublishing.com

Note: The information in this book is for educational purposes only and is not recommended as a means of diagnosing or treating an illness. All matters concerning physical and mental health should be supervised by a health practitioner knowledgeable in treating that particular condition. Neither the publisher nor the author directly or indirectly dispenses medical advice, nor do they prescribe any remedies or assume any responsibility for those who choose to treat themselves.

A cataloging-in-publication record for this book is available from the Library of Congress.

ISBN: 978-1-58054-456-6

Printed in the United States of America

Contents

Coenzyme Q10—The Essence of Energy

Every place your body makes energy, you'll find a natural compound called Coenzyme Q10—or CoQ10 for short. Its technical name is quite a mouthful—2,3 dimethoxy-5 methyl-6-decaprenyl benzoquinone, but it's more commonly known as ubiquinone or ubiquinol because it is ubiquitous—widespread and necessary for the basic functioning of all the cells in the body. In fact, that's how CoQ10 was initially discovered. Cellular biologists kept running into the same chemical every place where the body's demand for energy was high, leading them to suspect that the new compound played an important role in the body's cellular energy systems.

Coenzyme Q10 is a fat-soluble vitamin-like compound that is vital for activities related to energy metabolism. It's not actually a vitamin since all animals, including humans, can synthesize it. However, in many older adults or in those with particular health challenges the production of CoQ10 can decline. For these individuals, CoQ10 can be thought of as a "conditional vitamin," since boosting its level in the body through nutritional supplementation can have a positive impact on many aspects of health.

This illustration shows the general chemical form of Coenzyme Q in the Q3 configuration with three isoprenoid side chains rather than the ten chains of CoQ10.

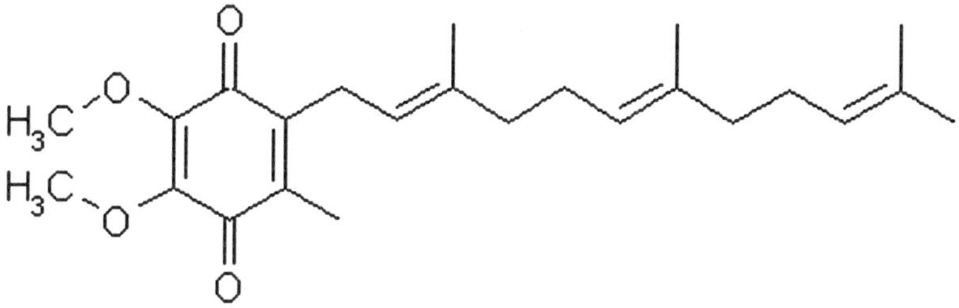

Enzymes and Coenzymes—Cellular Traffic Cops

The living body is a symphony of biochemical reactions that do everything from detoxifying our tissues and protecting us from infections to converting the food we eat into cellular energy. Enzymes are chemical compounds that enter the body through the foods we eat (especially raw fruits and vegetables, since heat destroys most enzymes) in addition to being generated by our own DNA (deoxyribonucleic acid).

Enzymes are cellular traffic cops that coordinate countless biochemical processes by speeding up the rate of particular reactions while effectively suppressing others. In the biological soup of the 100 trillion cells that make up the human body, enzymes are the primary agents of order and efficiency.

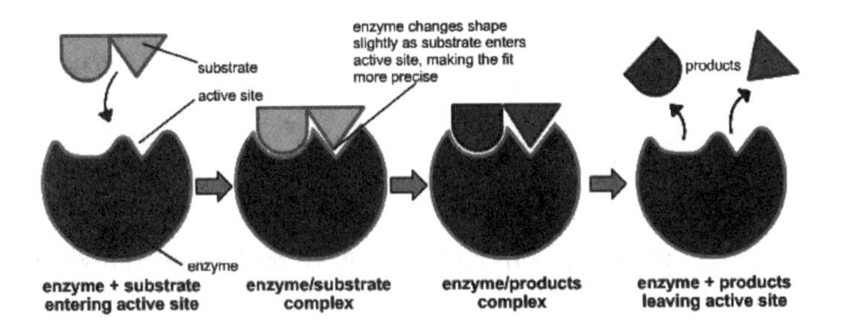

A coenzyme, like CoQ10, is a substance that enables an enzyme to do its job properly. CoQ10's primary activities involve our mitochondria—tiny cell-like structures called organelles that live within every one of our body's trillions of cells. Mitochondria are the body's power plants, converting nutrients into energy with the assistance of a variety of enzymes, coenzymes and minerals. CoQ10 is the key facilitator of at least three critical mitochondrial processes as well as other enzymes pathways vital to the creation of cellular energy.

Mighty Mitochondria—The Body's Battery Chargers

Throughout the body, a chemical called ATP, short for adenosine triphosphate, acts as a universal energy storage medium. Each molecule of ATP is actually a tiny battery holding two electrons. When the body needs energy to run for the bus, lift a bag of groceries, take a breath, make the heart beat, power the nervous system, zap a bacterium to prevent infection—anything at all—numerous ATP molecules pop apart, streaming out the electrons they're holding, much the same way that a battery floods electrons into the bulb of a flashlight to make a beam of light.

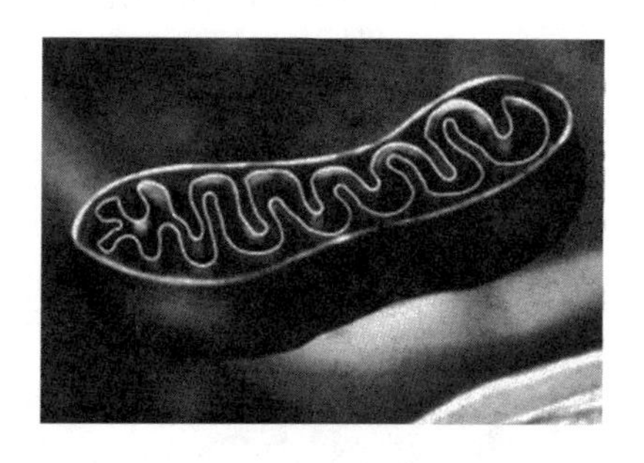

As an ATP molecule's first electron is pulled away to do the body's work, the molecule goes from being a fully charged ATP battery to becoming a partially charged ADP (adenosine diphosphate) battery because it only has one of its two electrons left. When ADP releases its final electron, it becomes a fully discharged battery called AMP (adenosine monophosphate).

In other words, as the body does work at the cellular level it fills up with discharged molecular batteries. These batteries need to somehow get recharged so the body can continue to function. Mitochondria are the body's cellular charging stations. They literally plug electrons back into AMP molecules to charge them into ADP molecules and then plug electrons back into ADP molecules to make fully charged ATP molecules.

CoQ10 is among the most essential nutrients for ensuring the proper and efficient operation of these mitochondrial charging stations. Naturally, cells that are called upon to do more work and expend more effort — such as muscle cells, heart cells, kidney cells and brain cells—have higher concentrations of mitochondria and therefore a greater need for CoQ10. It's easy to see that if a person's level of CoQ10 diminishes, either with age or illness, their ability to produce energy for every critical physical activity will diminish as well. This illustration shows the chemical structure of the ATP molecule.

The process of making ATP (and therefore cellular energy) from the food we eat has a great many parts, each of which depends on critical enzyme systems and the biochemical pathways they coordinate. CoQ10 plays a unique role in the final part of the process called the electron transport chain, in which the electrons that will be used to recharge the batteries are passed along.

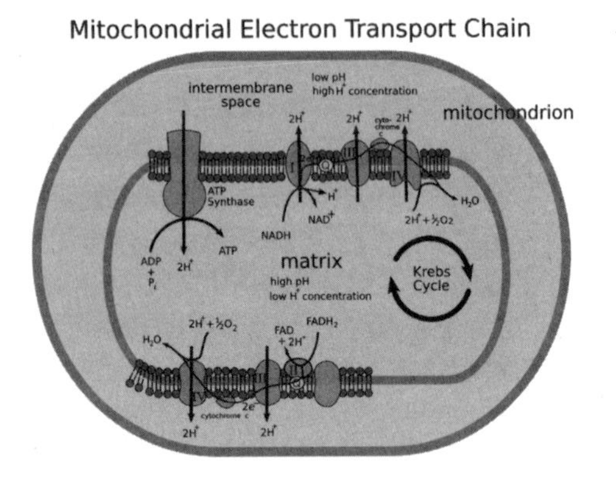

The electron transport chain, also known as the respiratory chain, is composed of the mitochondrial enzymes that transfer electrons from one complex to another, eventually resulting in the formation of ATP. This unique characteristic enables CoQ10 to move and transfer electrons between flavoproteins (respiratory enzymes) and cytochromes (iron-containing proteins important in cell respiration). Each pair of electrons processed by the electron transport interacts first with CoQ10. For this reason, the amount of energy the body can produce in this fashion actually depends on the availability of CoQ10, making it what scientists call the "rate-limiting element" of the mitochondrial respiratory chain. In other words, the amount of cellular energy we can produce is directly related to the amount of CoQ10 that's available.

CoQ10—A Synergistic Antioxidant

Along with serving as an electron and proton carrier in the mitochondrial respiratory chain, CoQ10 is a potent antioxidant. It inhibits lipid and protein peroxidation—breaking down to form more free radicals—and scavenges free radicals. CoQ10 is continually undergoing oxidation-reduction recycling. The reduced form readily gives up electrons to neutralize oxidants and displays its strongest antioxidant activity.

Some research indicates that CoQ10 prevents lipid peroxidation nearly as well as vitamin E. Other studies have shown CoQ10 to be more efficient in preventing LDL (low-density lipoproteins or the "bad" lipids associated with increased risk for cardiovascular disease) oxidation than vitamin E, lycopene, or ß-carotene. What's more, CoQ10 acts synergistically with vitamin E, producing a greater response than either would alone or than would be expected from the combination of both.

CoQ10 is the only fat-soluble antioxidant that is synthesized in our bodies. It is also a free radical scavenger and it is metabolized to ubiquinol, which prolongs the antioxidant effect of vitamin E. As we've observed, the highest amounts of it are found in the mitochondria of cells of organs with high energy requirements such as the heart muscle, liver, kidneys, and pancreas.

CoQ10 is a micronutrient that is naturally present in small amounts in a variety of foods. Organ meats such as heart, liver, and kidney, as well as beef, soybeans, soybean oil, sardines, mackerel, spinach, and peanuts, contain CoQ10. It was first isolated from the mitochondria of bovine hearts in 1957 at the University of Wisconsin, and identification of its chemical structure, which is quite similar to vitamin K, and synthesis were completed by 1958.

The average diet provides about 10 mg/day of CoQ10. Doses of 30–60 mg/day—roughly 1 mg/kg of body weight—are sufficient to prevent CoQ10 deficiency, however researchers have used considerably higher doses to achieve therapeutic effects. For optimal absorption of CoQ10, it should be taken with meals that contain fat.

Health Benefits of CoQ10

CoQ10 levels peak at about age nineteen and then decline as people age. Researchers have established average and normal blood and tissue levels of CoQ10, which range from 0.7–1.0 µg/mL (millionths of a gram in each thousandth of a liter).

CoQ10 levels may be low among people suffering from chronic diseases such as heart disease, cancer, diabetes, Parkinson's disease, muscular dystrophy, and HIV (human immunodeficiency virus, the virus that causes AIDS [acquired immunodeficiency syndrome]). CoQ10 has been used, recommended, or studied for numerous conditions, and scientific research has demonstrated that CoQ10 has a positive effect on a wide range of diseases and conditions. There is scientific evidence supporting the benefits of CoQ10 for health problems including:

- Amyotrophic lateral sclerosis (ALS)
- Angina—Preliminary evidence from small studies suggests that CoQ10 decreases angina and improves exercise tolerance.
- Alzheimer's disease—Some research has reported that CoQ10 may slow the progress of the dementia associated with Alzheimer's disease.
- Breast cancer
- Chronic kidney failure
- Congestive heart failure
- Friedreich's ataxia—Preliminary results of research suggest that CoQ10 may be beneficial in treating this genetic disorder that causes progressive deterioration of the nervous system.
- Gum disease—Decreased serum and gum tissue levels of CoQ10 have been documented in patients with periodontal (gum) disease. Several small studies suggest that CoQ10, taken orally or placed on the skin or gums, may benefit people suffering from periodontal disease by reducing pus, swelling, redness, pain and bleeding.
- Exercise tolerance and performance
- Heart attack
- Heart muscle damage in response to surgery, drug treatment or diabetes
- High blood pressure
- HIV/AIDS—There is some evidence that natural levels of CoQ10 in the body may be reduced in people with HIV/AIDS. CoQ10 may help to boost immunity and help to reduce the risk of contracting other infections.
- Huntington's disease
- Increasing sperm count—There is evidence that CoQ10 can help to increase sperm count and motility (the sperm's ability to swim or move).
- Migraine—There is some evidence to support the use of CoQ10 to prevent and treat migraines.
- Neuromuscular disorders—There is relatively strong support for the role of CoQ10 in treating muscular dystrophies and Parkinson's disease.

CoQ10 has also been used, and is being investigated, for its potential to help people suffering from:

- Anemia
- Asthma
- Bell's palsy
- Chronic fatigue—Some research suggests that a deficiency of CoQ10 may contribute to chronic fatigue syndrome.
- Diabetes—According to the National Center for Complementary and Alternative Medicine (NCCAM is one of the National Institutes of Health), there is insufficient evidence to evaluate CoQ10's effectiveness as a CAM therapy in diabetes. To date, CoQ10 has not been shown to affect blood glucose control. It may, however, help to prevent heart disease in people with diabetes.
- Hepatitis B
- Impaired immune function
- Liver disease
- MELAS (mitochondrial myopathy, encephalopathy, lactacidosis, stroke) syndrome
- Mitral valve prolapse
- Papillon-Le Fevre syndrome
- Sleep disorders
- Stomach ulcers
- Swelling

The Importance of Antioxidants

We've all heard that nutrients called antioxidants are important to our health. But if we look closely at the word *antioxidant* it seems like a paradox. After all, oxygen is necessary for life. Why should we want to oppose it with an antioxidant? Breathing oxygen is essential for life. If the brain is deprived of oxygen for more than a few minutes, irreparable damage occurs, followed by death. So why should the body need an abundance of antioxidants—chemicals that fight the effects of oxygen?

The answer to this question lies in the subtle distinction between oxygenation and oxidation, a truly profound difference that takes us back to our evolutionary roots. The fact that oxygen has two dimensions—on the one hand, it's an essential nutrient that we can't live without, and on the other, it's a ruthless destroyer that must be blocked and opposed—is known as the oxygen paradox.

How Our Cells Use Oxygen

With every inhalation—each breath of air we take—oxygen is drawn down into the lungs through a branching network of finer and finer tubes called bronchi. At the end of these bronchi the air inflates a spongy mass composed of billions of microscopic balloons called alveoli. Each alveolar bubble is wrapped in a network of tiny blood vessels so thin that the oxygen inside the alveoli passes directly into the blood. This oxygen-rich blood is then returned to the heart where it is pumped to all of the cells of the body.

Our cells, in turn, have the ability to absorb oxygen from the blood. One of the central biochemical processes within our cells takes this oxygen and delivers it, along with other nutrients including sugars and fats, to the energy-producing centers within the cell—specialized capsules called mitochondria. Mitochondria use these nutrients as fuel to produce ATP. As we've observed, ATP is the body's universal currency for storing and delivering energy where it's needed, for everything from running to catch a bus to processing the electrical signals in the nervous system that give rise to thought and awareness.

The Oxygen Paradox

The human body thrives in the presence of oxygen. Our cells burn sugars and fats in the presence of oxygen through an incredible efficient aerobic (oxygen utilizing) biochemical process called the Krebs cycle (also called the citric acid cycle or the tricyclic acid [TCA] cycle).

This constructive use of oxygen to generate cellular energy is called oxygenation. As long as oxygen is carefully directed into aerobic biochemical processes like the Krebs cycle, it's truly man's best friend. The diagram below provides a glimpse into the complexity of the Krebs cycle, which in turn, is just a small part of the network of chemical reactions that make up our human metabolism and internal biological terrain.

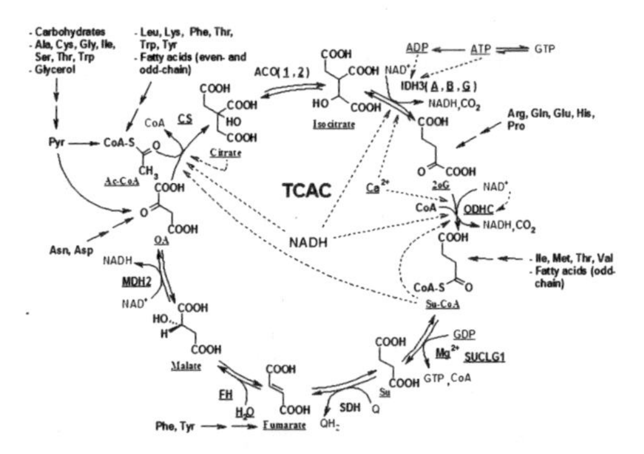

But oxygen has another action, one we see every day whenever we look at metallic objects that have been left out in the elements to weather and rust. Oxygen is a highly reactive, corrosive chemical. It can turn the strongest iron chain into a weak and crumbling wreck by corrupting it, one atom at a time, through a process known as oxidation.

Oxidation is a process in which oxygen (or certain other chemicals) attach themselves to other substances by stripping away their electrons. In the case of rust, oxygen attaches to iron to form oxide compounds that weaken and corrode the original structure.

Inside the living body, something quite similar can occur. Oxygen is capable of stripping an electron from another biochemical compound—effectively changing it into a positively charged ion since it has given up a negatively charged electron.

Sometimes this positive ion attaches to the oxygen or another negatively charged material. But on other occasions, the positively charged ion—which is now hungry for an electron so it can get back into electrical balance—strips an electron from a neighboring molecule, balancing itself but creating a new, imbalanced ion. The newly stripped ion can repeat this process and, like a line of dominos falling over, each one knocking down the next, a long chain of chemical changes can take place, each one damaging and degrading a previously balanced and functional biochemical substance.

Staying Between the Lines

As long as the oxygen within our bodies is directed into aerobic processes like the Krebs cycle, everything works perfectly. It's like cars speeding down the highway at sixty miles an hour. As long as they stay in their lanes, traffic flows along nicely and everyone travels smoothly and safely. But if the system of lanes breaks down and cars start weaving and moving around in every possible direction, then the situation immediately becomes dangerous and inefficient. It's the same way with oxygen, except that instead of painted lines to mark off lanes, our bodies have highly developed systems of antioxidants that try to keep atoms of oxygen traveling in the right direction—into the Krebs cycle—and prevent them from making trouble elsewhere in the body.

Some of the body's major antioxidant systems include glutathione peroxidase, superoxide dismutase, catalase and cytochrome P450. Some of these substances and the biochemical pathways they define are meant to keep oxygen heading in the right direction while others aim to keep it from heading off in the wrong direction.

Here's an analogy that might help put the two actions of oxygen—the positive action of oxygenation that feeds our cells and the negative action of oxidation that stresses them—into context:

The navy has a fleet of nuclear submarines. These remarkable ships are capable of remaining submerged in the ocean for months at a time as they travel around the world. The energy needed to power these huge machines comes from compact nuclear reactors that tap the energy of atomic nuclei to generate heat. In addition to driving the sub, this heat is used for everything on board—from scrubbing and recharging the air that the seamen breathe and purifying the water they drink to running the radar and information systems that keep them in touch and on track. If the nuclear generator on board were to fail, the lives of the crew would be imperiled.

On the other hand, the nuclear fuel needs to stay deep within the core of the reactor. Even though it produces the life-giving energy that every person on board depends on, the fuel itself is toxic. A few moments of direct exposure would create a lethal dose of radiation poisoning. The nuclear fuel must be kept within very specific "lanes"

—in this case, the physical shielding of the reactor core. It's the same with oxygen. It gives us life, but only when it is directed to and kept within the "core" of the biochemical processes that use oxygen to make energy. Like nuclear fuel escaping from the core, oxygen that escapes from the bounds of our antioxidant systems becomes very, very dangerous.

The Perils of Oxidative Stress

When oxygen slips outside of the lanes created in the body by antioxidants, it strips electrons from nearby substances, creating positively charged ions called free radicals. As we've described, one free radical, in an effort to find an electron to rebalance itself, damages another molecule, converting it into a free radical, and so forth. This domino-like series of damaging events is known as the oxidative stress cascade.

Oxidative damage causes many problems: First off, through the cascade process we've just described it can break down essential substances in the body, degrading them and even turning them into toxins and wastes that must be removed from the body, rather than helpful substances that serve it. Secondly, by locking up electrons, oxidatively damaged free radicals reduce the flow of electrons the body uses to create and direct energy at a cellular level.

Finally, the presence of free radicals in the body is a trigger for many inflammatory processes. Free radicals trigger the production of NF-κB (nuclear transcription factor kappa-B), which in turn signals the production of inflammatory cytokines—immune system substances produced by the body to fight cancer and other infections. While this can be helpful if triggered at the right time, when we are constantly exposed to oxidatively damaged free radicals, the overexpression of NF-κB and its related downstream cytokines can produce chronic inflammatory problems including arthritic symptoms, fibromyalgia, and even cardiovascular diseases.

Free Radicals, Oxidative Stress, and Disease

We know that oxidative stress is a harmful condition that can damage cells. It occurs when the balance of highly reactive, unstable molecules known as free radicals and antioxidants shifts in favor of the free radicals.

Free radicals are formed during a range of biochemical reactions and cellular functions including mitochondrial metabolism. Various conditions disrupt this balance by increasing the formation of free radicals in proportion to the available antioxidants. Oxidative stress describes the conditions, disorders, and diseases that lead to and increase the formation of free radicals, inflammation, infection, and cancer.

Free radicals are generally reactive oxygen or nitrogen species such as hydrogen peroxide, hydroxyl radical, nitric oxide, peroxynitrite, singlet oxygen, superoxide anion, and peroxyl radical. An example of a very potent free radical is peroxynitrite, which is one thousand times more potent as an oxidizing compound than hydrogen peroxide. Markers of peroxynitrite formation (such as nitrotyrosines or isoprostanes) have been found in many diseases including the brains of people suffering from neurodegenerative diseases such as Alzheimer's and Parkinson's disease as well as in people afflicted with chronic heart disease, liver disease, and inflammatory conditions. Inflammation, poor blood flow, degenerative diseases, and toxin exposures, among other conditions, all lead to oxidative stress.

Improving Heart Health and Preventing Cardiovascular Disease

CoQ10 is highly concentrated in heart muscle cells because these cells have high energy requirements. It helps cells convert nutrients into energy and is a powerful antioxidant—attacking substances in the body that may damage healthy cells. These actions have been found to support healthy heart function and may play a role in helping to prevent, treat, and reverse some forms of cardiovascular disease.

The bulk of the research conducted about CoQ10 has focused on the role of CoQ10 in heart health. Research that began during the late 1970s and 1980s confirmed that animals and people with cardiovascular disease had lower levels, even deficiencies, of CoQ10 in their heart muscle tissue. At the conclusion of a long-term study of the effectiveness of using CoQ10 to treat a variety of cardiovascular diseases, researchers from the University of Texas concluded that, "CoQ10 is a safe and effective adjunctive treatment for a broad range of cardiovascular diseases, producing gratifying clinical responses while easing the medical and financial burden of multidrug therapy."

Several forms of cardiovascular disease including congestive heart failure (a disorder in which the heart loses its ability to pump efficiently) have been strongly correlated with significantly low blood and tissue levels of CoQ10. The severity of heart failure directly relates to the severity of CoQ10 deficiency. Animal studies have shown that when given CoQ10, the animals' hearts were protected from damage by acute oxidative stress and the functional damage caused by a compromised blood supply.

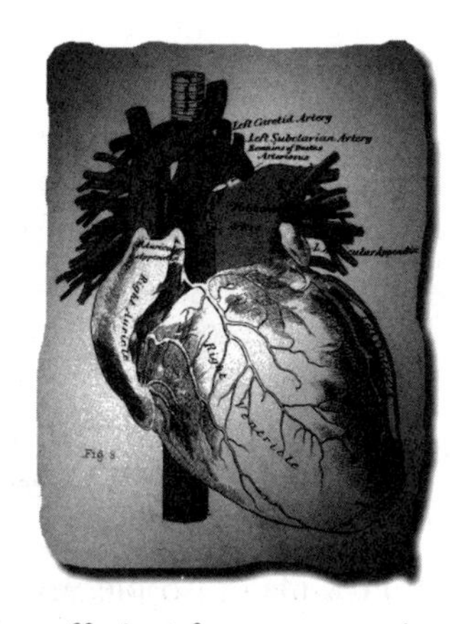

CoQ10 has proven useful for people suffering from congestive heart failure, cardiomyopathy (weakening of the heart muscle), angina (pain resulting from lack of oxygen to the heart muscle), heart attack (damage to the heart muscle due to insufficient blood supply), and hypertension (high blood pressure). CoQ10 also may offer some protection against damage to the heart muscle caused by diabetes, cardiac surgery—coronary artery bypass graft surgery and heart transplants—and cancer chemotherapy.

In a small randomized double-blind placebo-controlled study, researchers gave 60 mg of CoQ10 twice daily to subjects who had suffered a heart attack within seventy-two hours and compared how they fared to subjects given a placebo. After one year, total cardiac events, including second heart attacks, stroke, and cardiac deaths were significantly lower among the subjects given CoQ10.

Two recent studies found that giving CoQ10 to people who had suffered cardiac arrest and survived cardiopulmonary resuscitation (CPR) helped to protect their brains from injury. Many people resuscitated using CPR suffer some form of brain damage, largely in response to the temporary loss of blood to the brain but also after the return of blood flow to the brain. Brain injury may be compounded by inflammation, problems with the function of mitochondria, oxidative stress, altered signal transduction (the conversion of a signal from outside the cell to a functional change within the cell such as cell division or glucose uptake), and programmed cell death. The studies found that combining CoQ10 with mild hypothermia—cooling the body's core temperature to 35 degrees Celsius for twenty-four hours immediately after CPR—improved survival and neurological functioning of survivors.

Another randomized double-blind placebo-controlled study considered whether CoQ10 would lower blood lipids—total cholesterol and LDL (low-density lipoprotein, or "bad") cholesterol —in subjects with elevated blood lipids and diagnosed heart disease. They found that blood lipids decreased significantly—a nearly 25 percent reduction—among the subjects taking CoQ10. The subjects taking CoQ10 also had reduced blood glucose levels and reductions in blood markers for oxidative stress.

CoQ10 also may be useful for relieving the muscle pain that afflicts many people who take lipid-lowering drugs called statins. Some researchers have hypothesized that inhibition of CoQ10 synthesis explains some of the commonly reported side effects of statins, especially exercise intolerance and muscle pain. Several studies have shown that statins reduce CoQ10 levels and that taking CoQ10 in no way compromises the statins' lipid-lowering effects.

In fact, because of the antioxidant effects of CoQ10's metabolite, ubiquinol, another potential benefit of taking CoQ10 along with lipid-lowering drugs may be decreased oxidation of LDL, resulting in a reduction of this lipid that is linked to development of atherosclerosis—clogging, narrowing, and hardening of the large arteries and medium-sized blood vessels that can lead to heart attack and stroke. Another proposed explanation of the role of CoQ10 in preventing atherosclerosis is its observed ability to decrease the expression of cell adhesion molecules that recruit monocytes to blood vessel walls—a process that is key to the development of atherosclerosis.

Considerable research has found favorable effects of CoQ10 on other measures of cardiac function including ejection fraction (the measurement of the blood pumped out of the ventricles of the heart), exercise tolerance, cardiac output (the amount of blood that goes through the circulatory system in one minute), and stroke volume (the amount of blood ejected by one ventricle of the heart with each beat). Similarly, CoQ10 supplementation has been shown to be safe with a very low incidence of adverse effects. The results of several small studies also indicate that CoQ10 may be helpful in lowering blood pressure.

Cancer

An increasing amount of laboratory research and epidemiological (population study) evidence indicates the involvement of oxygen-derived free radicals in the development of cancer. Free radicals have been found to be involved in both initiation and promotion of carcinogenesis—the process by which normal cells are transformed into cancer cells.

According to the National Cancer Institute (NCI is one of the National Institutes of Health), CoQ10's antioxidant action, ability to support immune function, and role in cellular energy are cited as the probable sources of its anticancer activity. The body not only uses it for cell growth but also to protect cells from damage that could lead to cancer.

CoQ10 confers this protection because as a powerful antioxidant, it protects cells from free radicals, which can damage DNA. Damage to DNA is associated with some kinds of cancer. So, by protecting the body from free radicals, antioxidants like CoQ10 may help to defend the body from cancer.

Animal studies also have found that coenzyme Q10 stimulates the immune system and increases resistance to disease. One study found that the combination of tamoxifen, a potent antioxidant and a non-steroidal anti-estrogen drug frequently used in the chemotherapy and chemoprevention of breast cancer, and CoQ10 served to prevent cancer cell proliferation.

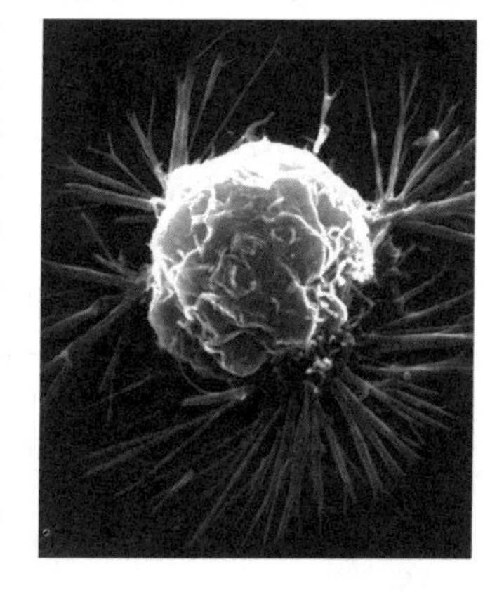

In a small study, a group of breast cancer patients was treated with antioxidants, fatty acids, and 90 mg of

CoQ10 per day. Six of the thirty-two patients showed partial regression of their tumors. In one of these six subjects, the dose of CoQ10 was increased to 390 mg. After one month, the tumor could no longer be felt by manual examination, and in another month mammography confirmed the absence of tumor.

Encouraged by this result, another subject with a breast tumor who had undergone non-radical surgery and had verified residual tumor was then treated with 300 mg CoQ10 per day. After three months, the subject was in excellent health and there was no residual tumor tissue. The investigators hypothesized that the bioenergetic activity of CoQ10, expressed as hematological or immunological activity, may be the primary but not the sole molecular mechanism prompting the regression of breast cancer.

Some other small studies suggest that CoQ10 may help to prevent and treat cancer as well as reduce chemotherapy-related damage, especially heart muscle damage. Many anticancer drugs are known to generate high levels of oxidative stress and nearly all anticancer drugs generate some free radicals as they induce apoptosis (programmed cell death) in cancer cells.

Although doxorubicin, the anticancer drug implicated in heart muscle damage, does not itself produce a significant amount of free radicals, it disrupts the electron transport system and harms heart cells by generating hydroxyl radicals in mitochondria of cardiac cells. This in turn inhibits CoQ10–dependent enzymes, disrupting mitochondrial energetics and producing cardiotoxicity, which is evidenced by arrhythmias (abnormal heart rhythms that can cause the heart to pump less effectively) and reduced ejection fraction. Although clinical studies have shown that simply giving patients antioxidants does not prevent the doxorubicin-induced cardiotoxicity, giving CoQ10 does prevent it, without interfering with the drug's anticancer activities.

Other research suggests that deficiencies of CoQ10 may contribute to the development of some cancers. Several studies in women with breast cancer found reduced levels of CoQ10 in diseased breast tissue and blood. Preliminary research indicates that increasing CoQ10 levels may be helpful in the prevention and treatment of cancer, however, to date there are no published results of clinical trials of CoQ10 as prevention or treatment for cancer.

Neuromuscular and Neurodegenerative Diseases

Amyotrophic Lateral Sclerosis (ALS)

Amyotrophic lateral sclerosis, or ALS, is a disease of the motor nerve cells in the brain and spinal cord that causes progressive loss of motor control. The French refer to ALS as *maladie de Charcot*, named after French physician Jean-Martin Charcot, who first wrote about ALS in 1869. In the United States, it's often referred to as "Lou Gehrig's disease," because the famous Yankees baseball player valiantly and publicly battled the disease during the late 1930s, succumbing to it in 1941 at age thirty-eight.

According to the ALS Association, just one out of every 100,000 people will develop ALS and most people who develop the disease are between forty and seventy years old. Symptoms usually appear after age fifty, beginning with progressive loss of muscle strength and coordination that eventually interferes with the ability to perform routine activities, and also may impair vital bodily functions such as breathing and swallowing. In about 10 percent of cases, ALS is caused by a genetic defect. In other cases, the cause of the nerve deterioration is unknown.

There is no known cure for ALS. Drugs, such as riluzole may prolong life, but do not reverse or stop disease progression. Treatment aims to control symptoms such as the characteristic spasticity that interferes with activities of daily living or impaired ability to swallow saliva. Physical therapy, rehabilitation, and the use of durable medical equipment such as braces or a wheelchair are often used to maximize muscle function and general health.

As the disease progresses, more muscle groups are affected and patients become progressively incapacitated. Although ALS is not associated with cognitive impairment, some patients suffer from emotional lability, finding it difficult to control crying or laughing. Nearly half of all people with ALS live at least three years after they have been diagnosed. Twenty percent live five years or more, and up to 10 percent will survive more than ten years.

One of the most prominent long-term survivors of ALS is the world-renowned physicist Stephen Hawking. Hawking has been living with ALS disease for more than forty years—ever since his diagnosis at age twenty-one.

Alzheimer's Disease

Alzheimer's disease (AD) is a progressive, degenerative disease that affects the brain and results in severely impaired memory, thinking, and behavior. It is the fourth leading cause of death in adults, and the incidence of the disease rises with age. AD affects an estimated four million American adults and is the most common form of dementia, or loss of intellectual function. The U.S. Department of Health and Human Services estimates that in 2004, 4.5 million Americans suffered from AD.

The German physician Alois Alzheimer first described the disease in 1907 after he had cared for a patient with an unusual mental illness. Dr. Alzheimer observed anatomic changes in his patient's brain and described them as abnormal clumps and tangled bundles of fibers. Nearly a century later, these abnormal findings, now described as amyloid plaques and neurofibrillary tangles, along with abnormal clusters of proteins in the brain, are recognized as the characteristic markers of AD.

The following images show positron emission tomography (PET) scans of a normal brain (top) and a brain ravaged by advanced Alzheimer's disease (bottom). These images can be found on the Web site of the National Institute of Aging (NIA is one of the National Institutes of Health).

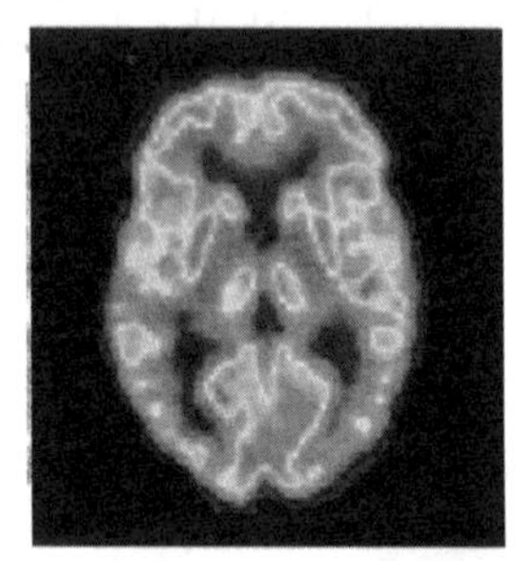

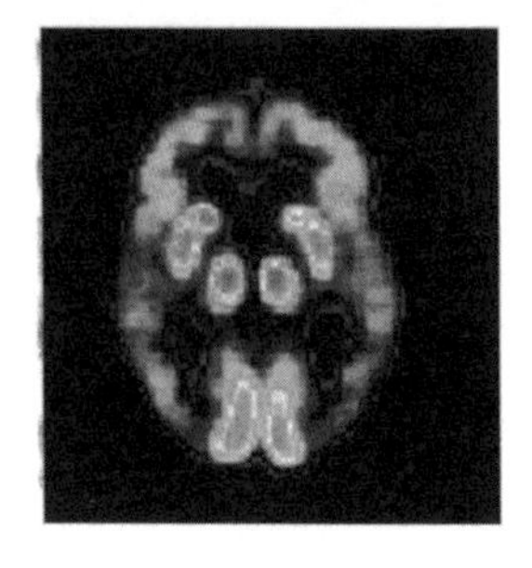

AD begins slowly. The symptoms include difficulty with memory and a loss of cognitive function—memory and intellectual abilities. The patient with AD also may experience confusion; language problems, such as trouble finding words; impaired judgment; disorientation in place and time; and changes in mood, behavior, and personality. How quickly these changes occur varies from person to person, but eventually the disease leaves its victims unable to care for themselves. Toward the ends of their lives, patients with AD require care twenty-four hours a day. They no longer recognize family members or themselves, and they need help with such daily activities as eating, dressing, and bathing. Eventually, they may become incontinent, blind, and unable to communicate.

There is no cure or prevention for AD, and medical treatment focuses on managing symptoms. Medication can reduce some of the symptoms, such as agitation, anxiety, unpredictable behavior, and depression. Physical exercise and good nutrition are important, as is a calm and highly structured environment. The object is to help the patient with AD maintain as much comfort, normalcy, and dignity as possible.

The disease knows no social or economic boundaries and affects men and women almost equally. While 90 percent of AD victims are over age sixty-five, AD can strike as early as the thirties, forties, and fifties. Most patients are cared for at home as long as possible, a situation that can be emotionally and physically devastating for the affected individuals and their families.

The Alzheimer's Association asserts that if a cure or prevention is not found by 2050, the number of Americans with AD is estimated to range from 11.3 million to 16 million, with a middle estimate of 13.2 million. On the other hand, discovery of a treatment that could delay the onset of AD by five years could reduce the number of individuals with the disease by nearly 50 percent after fifty years.

Parkinson's Disease

Parkinsonism refers not to a particular disease but to a condition marked by a characteristic set of symptoms that, according to the American Parkinson Disease Association, afflicted about one million people in the United States in 2006. Men and women are affected equally and the probability of developing Parkinson's disease (PD) increases with advancing age. PD usually strikes people in their sixth and seventh decades—the average age of symptom onset is 62.4 years, but up to 10 percent of patients are afflicted by age forty.

PD is caused by the progressive deterioration of about half a million brain cells in the portion of the brain that controls certain types of muscle movement. These cells secrete dopamine, a neurotransmitter (chemical messenger). Dopamine's function is to allow nerve impulses to move smoothly from one nerve cell to another. These nerve cells, in turn, transmit messages to the muscles of the body to begin movement. When the normal supply of dopamine is reduced, the messages are not correctly sent, and the symptoms of PD appear.

The four early warning signs of PD are tremors, muscle stiffness, unusual slowness, and a stooped posture. Medications can control initial symptoms, but as time goes on they become less effective. As the disease worsens, patients develop tremors, causing them to fall or jerk uncontrollably. (The jerky body movements that patients with PD experience are known as dyskinesias.) At other times, rigidity sets in, rendering patients unable to move. About one-third of patients also develop dementia.

Management of PD includes drug therapy and daily exercise. Exercise often can reduce the rigidity of muscles, prevent weakness, and improve the ability to walk. The main goal of drug treatment is to restore the chemical balance between dopamine and another neurotransmitter, acetylcholine. The standard treatment for most patients is levodopa (L-dopa), which was first approved for use in 1970. L-dopa is a compound that the body converts into dopamine to replace it in the body and help alleviate symptoms. (Without dopamine, signals from the brain cannot be transmitted properly to the body, and movement is impaired.) Treatment with L-dopa does not, however, slow the progressive course of the disease or even delay the changes in the brain that PD produces, and it may produce some unpleasant side effects because of its change to dopamine before reaching the brain. Simultaneously administering substances that inhibit this change allows a higher concentration of levodopa to reach the brain and also considerably decreases the side effects.

Five classes of drugs are used to treat the symptoms of PD. Anticholinergics work to relieve tremor and rigidity. COMT (catechol-O-methyltransferase) inhibitors act by prolonging the effectiveness of a dose of levodopa by preventing its breakdown. MAO-B inhibitors slow the breakdown of dopamine in the brain. Amantadine has demonstrated effectiveness in reducing dyskinesias.

Huntington's Disease

Named for an American physician, George Sumner Huntington (1850–1916), Huntington's disease (HD) is an inherited, progressive brain disorder. It causes the degeneration of cells in the basal ganglia, a pair of nerve clusters deep in the brain that affect both the body and the mind. Once considered rare, HD is now recognized as one of the more common hereditary diseases. It affects about 30,000 Americans; another 150,000 are at a 50 percent risk of inheriting it from an affected parent. Estimates of its prevalence are about one in 10,000 people.

A single dominant gene causes HD. The gene mutation that produces HD was mapped to chromosome 4 in 1983 and cloned in 1993. The mutation is in the DNA that codes for the protein huntingtin. The number of repeated triplets of nucleotides, cytosine (C), thymine (T), and guanine (G), known as CTG, is inversely related to the age when the individual first experiences symptoms—the more repeated triplets, the younger the age of onset of the disease. Like myotonic dystrophy, in which the symptoms of the disease often increase in severity from one generation to the next, the unstable triplet repeat sequence can lengthen from one generation to the next, with a resultant decrease in the age when symptoms first appear.

HD does not usually strike until mid-adulthood, between ages thirty and fifty, although there is a juvenile form that can affect children and adolescents. Early symptoms, such as forgetfulness, a lack of muscle coordination, or a loss of balance, are often ignored, delaying the diagnosis. The disease gradually takes its toll over a ten-to-twenty-five-year period.

Within a few years, characteristic involuntary movement (often termed "chorea") of the body, limbs, and facial muscles appears. As HD progresses, speech becomes slurred and swallowing becomes difficult. The patients' cognitive abilities decline and there are distinct personality changes—depression and withdrawal, sometimes countered with euphoria. Eventually, nearly all patients must be institutionalized, and they usually die as a result of choking or infections.

Migraine

Migraine is a chronic neurological disease characterized by episodic attacks of severe, often disabling headaches and associated symptoms such as nausea, vomiting, and extreme sensitivity to light and sound. The quality of the headache is often described as pulsing or throbbing; it is exacerbated by physical activity and may occur on just one side of the head. An estimated 11 percent of adults—more than 28 million Americans—suffer from migraines.

Many migraine sufferers find their headaches are preceded or accompanied by a kind warning, known as an aura, which may be experienced as flashes of light, blind spots, or tingling in the extremities. Others can predict that they will have a migraine days or hours before it occurs because they experience specific feelings or symptoms such as a burst of energy, intense thirst, craving for sweets, sleepiness, or depression.

Untreated, migraines can persist as long as seventy-two hours, but the frequency with which they occur varies widely—some people experience them as often as once a week, others have just one a year. Some migraine sufferers are able to identify triggers—specific foods, high-intensity exercise, emotional stress, or hormonal fluctuations—that seem to precipitate or aggravate migraines.

Muscular Dystrophy

Muscular dystrophy (MD) is a term that applies to a group of hereditary muscle-destroying disorders. According to the Muscular Dystrophy Association (MDA) in 2006, some type of MD affected about one million Americans. Each variant of the disease is caused by defects in the genes that play important roles in the growth and development of muscles. In MD, the proteins produced by the defective genes are abnormal, causing the muscles to waste away. Unable to function properly, the muscle cells die and are replaced by fat and connective tissue. The symptoms of MD may not be noticed until as much as 50 percent of the muscle tissue has been affected.

All of the various disorders labeled MD cause progressive weakening and wasting of muscle tissues. They vary, however, in the usual age at the onset of symptoms, rate of progression, and initial group of muscles affected. The most common type, Duchenne MD, affects young boys, who show symptoms in early childhood and usually die from respiratory weakness or damage to the heart before adulthood. The gene is passed from the mother to her children. Girls who inherit the defective gene generally do not manifest symptoms—they become carriers of the defective genes, and their children have a 50 percent chance of inheriting the disease. Other forms of MD appear later in life and are usually not fatal.

In 1992 scientists discovered the defect in the gene that causes myotonic dystrophy, the most common form of MD. In people with this disorder, a segment of the gene is enlarged and unstable. This finding helps physicians more accurately diagnose myotonic dystrophy. Researchers since have identified genes linked to other types of MD, including Duchenne MD, Becker MD, limb-girdle MD, and Emery-Dreifuss MD.

CoQ10 May Slow the Progress of Some Diseases

The high metabolic load of nerve cells, especially motor neurons, and their dependence on oxidative phosphorylation—the formation of ATP from the energy released by the oxidation of various substrates, especially the organic acids involved in the Krebs cycle—may make them particularly vulnerable to the loss of mitochondrial function. Since CoQ10 is an antioxidant and an essential mitochondrial cofactor that facilitates electron transfer in the electron transport chain, it has been tested as treatment for neurodegenerative conditions in which mitochondrial dysfunction has been implicated.

Modest improvements in symptoms and slowing of the progression of neurological disorders including Friedreich's ataxia and Alzheimer's disease after CoQ10 use have been reported in the literature. Friedreich's ataxia is a genetic disorder resulting in progressive deterioration of the nervous system that causes an inability to coordinate voluntary muscle movement. It usually begins between ages ten and thirteen, and starts with an unsteadiness in the legs. Over the course of eight to ten years, the affected individual loses the ability to walk unassisted. Alzheimer's disease is a progressive, neurodegenerative disease characterized by loss of function and death of nerve cells in several areas of the brain, resulting in loss of mental functions such as memory and learning. It is the most common cause of dementia.

Amyotrophic Lateral Sclerosis

Deficiencies in CoQ10 and resulting mitochondrial dysfunction have been implicated as possible causes and contributors to a variety of neuromuscular diseases including ALS, HD, PD, and some forms of muscular dystrophy.

Investigators at Massachusetts General Hospital and Harvard Medical School hypothesized that because CoQ10 is a cofactor in mitochondrial electron transfer it might improve the mitochondrial dysfunction that contributes to cell death in people suffering from ALS. Since earlier animal studies suggested that high doses of CoQ10 were beneficial for neurodegenerative disorders, the investigators sought to determine whether very high doses (up to 3,000 mg per day) would be safely tolerated. They found that the highest doses of CoQ10 were safe and well tolerated and that there was no statistically significant difference between blood plasma levels of CoQ10 at 2,400 mg and 3,000 mg doses—plasma levels appeared to plateau at 2,400 mg.

Parkinson's and Huntington's Diseases

CoQ10 has been shown to benefit symptoms of neurodegeneration. A randomized double-blind placebo-controlled multicenter study, led by researchers at the University of California, San Diego, School of Medicine, looked at a total of eighty PD patients at ten centers across the country to determine if coenzyme Q10 is safe and if it can slow the rate of functional decline.

All of the subjects who took part in the new study had the three primary features of PD—tremor, stiffness, and slowed movements—and had been diagnosed with the disease within five years of the time they were enrolled in the study. After an initial screening and baseline blood tests, the subjects were randomly divided into four groups. Three of the groups received CoQ10 at three different doses (300 mg/day, 600 mg/day, and 1,200 mg/day), along with vitamin E, while a fourth group received a placebo that contained vitamin E alone. Each subject received a clinical evaluation one month later and every four months for a total of sixteen months.

During the study period, the group that received the largest dose of CoQ10 (1,200 mg/day) had 44 percent less decline in mental function, motor (movement) function, and ability to carry out activities of daily living, such as feeding or dressing themselves. The greatest effect observed by the investigators was an improvement in the ability to perform the activities of daily living.

The groups that received 300 mg/day and 600 mg/day developed slightly less disability than the placebo group, but the effects were less than those in the group that received the highest dosage of CoQ10. The groups that received CoQ10 also had significant increases in the level of CoQ10 in their blood and a significant increase in energy-producing reactions within their mitochondria.

The results of this study suggest that doses of CoQ10 as high as 1,200 mg/day are safe and may be more effective than lower doses. Another study of twenty-eight patients with Parkinson's disease reported mild improvement of symptoms with a daily dose of 360 mg of CoQ10. Although there was little reported improvement in motor symptoms, subjects who received CoQ10 performed better than those receiving placebo on a test of color vision.

CoQ10 has also demonstrated benefits for people suffering from Huntington's disease, an inherited condition characterized by abnormal body movements, dementia, and psychiatric problems. Researchers conducted a randomized controlled trial that gave subjects either CoQ10 (600 mg/day), remacemide (a drug that blocks a receptor for glutamate, a neurotransmitter involved in cell death), or a combination of remacemide and CoQ10 for thirty months. Although their results were not statistically significant, the researchers found that the CoQ10 measurably slowed functional declines in the HD patients while the remacemide did not. Specifically, they observed a trend toward improvement in color naming, word reading, and attention.

Migraine

Research studies, including DNA analyses, indicate that some migraines may be caused by dysfunctional mitochondria resulting in impaired oxygen metabolism. This finding prompted researchers to think that there might be a role for CoQ10 in migraine prevention since it is an essential element of the mitochondrial electron-transport chain. Several studies have reported that CoQ10 was more effective than placebo in terms of preventing migraines.

In one study, thirty-two subjects who suffer from migraines were given 150 mg per day of CoQ10 for three months. All but one of the subjects had more than a 50 percent reduction in the number of days they suffered from migraine headaches. The average number of days the subjects had migraines dropped from 7.34 before treatment to 2.95 days after three months of treatment. This decrease was statistically significant. Similarly, the frequency of migraines decreased from 4.85 before treatment to 2.81 by the end of the study period.

Another randomized double-blind placebo-controlled study compared CoQ10 to placebo for migraine prevention. Swiss researchers theorized that migraines may be caused by a decrease in mitochondrial energy reserve, and that CoQ10 would serve as an energy boost in the brain. They gave migraine sufferers 100 mg of CoQ10 three times per day and found that the CoQ10 was significantly better than placebo at reducing the frequency of migraines and reducing the associated gastrointestinal symptoms—nausea and vomiting.

Approximately half of those who took CoQ10 had a 50 percent response rate—a reduction in the frequency and severity of attacks—during the three-month study, while this occurred in just 14 percent of those taking a placebo. The number of migraine attacks per month was reduced in the treatment group from 4.4 to 3.2, with no change in the placebo group. The CoQ10 seemed to take effect after about one month, and the subjects received the maximum benefit from it at three months.

Ongoing CoQ10 Research

Nationally and internationally, researchers in the public sector, government, and university teaching hospitals are actively studying the potential of CoQ10 to prevent and treat various diseases. The clinical trials described here are just examples of the many studies presently underway.

The National Institute of Neurological Disorders and Stroke (NINDS is one of the National Institutes of Health) is conducting a randomized double-blind placebo-controlled clinical trial of high doses of CoQ10 to treat 185 people with amyotrophic lateral sclerosis (ALS). Subjects are being monitored, and their forced vital capacity (measures air flow into and out of the lungs when the subject is exhaling with maximum speed and effort), fatigue severity, and abilities to perform the activities of daily living are being assessed over a nine-month period. The goal of this study is to compare two doses of CoQ10 to determine safety and efficacy and to select the preferred dose for a larger clinical study involving more subjects.

The Cooperative International Neuromuscular Research Group (sponsored by the U.S. Department of Defense) is conducting a clinical trial to help determine if CoQ10 and prednisone (a synthetic steroid similar to cortisone), alone and in combination, decrease the decline in cardiopulmonary and skeletal muscle function that occurs in patients during the wheelchair-confined phase of Duchenne muscular dystrophy (DMD is the most common form of muscular dystrophy affecting 1 in 3,500 male births worldwide).

Even with markedly enhanced understanding of the disorder since the discovery of the causative gene and its product dystrophin in 1987, current therapy remains largely supportive. Improvement in the treatment of DMD depends on the development of better therapies. Affected boys develop symptoms at three to five years of age with leg weakness that impairs mobility, ability to get up from a squat, and impairs a normal ability to run. By eight years of age, some affected boys begin to lose the ability to walk and resort to a wheelchair for mobility. This shift from walking to being wheelchair-dependent occurs in all boys with a diagnosis of DMD by age twelve. In this study, ten-to-fifteen-year-old boys with DMD will be treated and followed for a twelve-month investigation period.

References

Beal, M. F., C. W. Shults. "Effects of Coenzyme Q10 in Huntington's Disease and Early Parkinson's Disease." *Biofactors*. 2003; 18: 153–61.

Conklin, K. A. "Cancer Chemotherapy and Antioxidants." *J Nutr*. 2004; 134: 3201S–3204S.

Damian M. S., et al. "Coenzyme Q10 Combined with Mild Hypothermia after Cardiac Arrest." *Circulation*. 2004; 110: 3011–16.

Ferrante, K. L., et al. "Tolerance of High-dose (3,000 mg/day) Coenzyme Q10 in ALS." *Neurology*. 2005; 65(11): 1834–36.

Folkers, K., S. Vadhanavikit, S. A. Mortensen. "Biochemical Rationale and Myocardial Tissue Data on the Effective Therapy of Cardiomyopathy with Coenzyme Q10." *Proc. Natl. Acad. Sci.* 1985; 82(3), 901–04.

Langsjoen, H. et al. "Usefulness of Coenzyme Q10 in Clinical Cardiology: A Long-term Study." *Mol Aspects Med.* 1994; 15 Suppls. 165–75.

Lockwood, K., S. Moesgaard, K. Folkers. "Partial and complete regression of breast cancer in patients in relation to dosage of coenzyme Q10." *Biochem Biophys Res Commun.* 1994; 199 (3): 1504–08.

Miller, K. L., et al. "Complementary and alternative medicine in cardiovascular disease: A review of biologically based approaches." *Am Heart J.* 2004; 147: 410–11.

Mortensen, S. A., S. Vadhanavikit, K. Folkers. "Deficiency of coenzyme Q10 in myocardial failure." *Drugs Exptl. Clin. Res.* 1984; X(7) 497–502.

Muller, T., et al. "Coenzyme Q10 supplementation provides mild symptomatic benefit in patients with Parkinson's disease." *Neurosci Lett* 2003; 341: 201–04.

Perumal, S. S., P. Shanthi, P. Sachdanandam. "Combined efficacy of tamoxifen and coenzyme Q10 on the status of lipid peroxidation and antioxidants in DMBA-induced breast cancer." *Mol Cell Biochem.* 2005; 273 (1–2): 151–60.

Rapoport, A. M., M. E. Bigal. "Migraine Prevention Therapy." *Neurol Aci.* 2005; 26: S111–S120.

Sandor, P. S., et al. "Efficacy of coenzyme Q10 in migraine prophylaxis: A randomized controlled trial." *Neurology.* 2005; 64: 713–15.

Shults, C. W., et al. "Effects of coenzyme Q10 in early Parkinson disease: evidence of slowing of the functional decline." *Arch Neurol.* 2002; 59: 1541–50.

Siciliano, G., et al. "Coenzyme Q10, exercise lactate and CTG trinucleotide expansion in myotonic dystrophy." *Brain Res Bull.* 2001; 1; 56 (3–4): 405–10.

Singh, R. B., et al. "Effect of coenzyme Q10 on risk of atherosclerosis in patients with recent myocardial infarction." *Mol Cell Biochem.* 2003; 246 (1–2): 75–82.

Singh, R. B., M. A. Niaz. "Serum concentration of lipoprotein(a) decreases on treatment with hydrosoluble coenzyme Q10 in patients with coronary artery disease: discovery of a new role." *Int J Cardiol.* 1999; 68(1): 23–09.

Strong, M. J., G. L. Pattee. "Creatine and coenzyme Q in the treatment of ALS. *Amyotrophic Lateral Sclerosis & Other Motor Neuron Disorders.* 2000; Suppl 4; 1(5): 17–21.

Tran, M. T., et al. "Role of coenzyme Q10 in chronic heart failure, angina, and hypertension." *Pharmacotherapy.* 2001; 21 (7): 797–806.